THE BOOK of FIRE

Second Edition

William H. Cottrell Jr.

With a Foreword by Jane Kapler Smith

2004
Mountain Press Publishing Company
Missoula, Montana
IN COOPERATION WITH THE NATIONAL PARK FOUNDATION

To Mom and Dad,
whose parenting permitted me to be
more curious than cautious
about life's mysteries.

The foreword was written and prepared by a U.S. Government employee
on official time, and therefore it is in the public domain
and not subject to copyright.

The Murphy-Phoenix Company generously granted the financial
support that made the production of the first edition possible.
In publishing, a colored picture is worth a thousand words.

Cover and Design by Kim Ericsson

Library of Congress Cataloging-in-Publication Data
Cottrell, William H.
    The book of fire / William H. Cottrell, Jr.— 2nd ed.
      p. cm.
  Includes bibliographical references (p.   ).
   ISBN 0-87842-491-1 (pbk. : alk. paper)
  1. Forest fires. 2. Fire. I. Title.
  SD421.C714 2004
  634.9'618—dc22
                                    2003020773

PRINTED IN HONG KONG BY MANTEC PRODUCTION COMPANY

Mountain Press Publishing Company
P.O. Box 2399 • Missoula, MT 59806

Contents

Foreword

Every child I have met is either terrified of fire or wants to play with it. This is true of most adults too. I have watched senior citizens, working in a laboratory for the first time ever, light single matches over and over to observe the heat dispersal patterns and exclaim, "Well, isn't that something!"

Since the late 1980s, extended drought and increased development near wildlands have contributed to hundreds of large, severe fires and huge property losses. Our interest has grown, and our questions have become urgent. What is the fire going to do? Will it damage the land? Will it endanger firefighters? Destroy homes? How will the smoke affect our health? When land managers actually start fires to change the landscape—to benefit plants and animals, or to reduce the debris on the ground—our questions only increase. Will the fire be too severe? Will it escape control? We need fire and yet fear it, so we need greater understanding. That is what this book provides.

We could go directly to scientists for knowledge, but they may start with scientific laws and calculations. *The Book of Fire* starts with what an average person knows: matches, candles, even our own breath. Furthermore, it is fundamentally a picture book rather than a word book. The short paragraphs focus on explaining the abundant, vivid illustrations; the text doesn't digress. The book begins with the basic chemistry of fire, but the explanation is not abstract; it is based on tangible elements of daily life: water, temperature, and the flame from a burning candle. The book discusses smoke from wildland fires by relating it to diagrams of smoke from scorched toast and cigarettes.

The book describes severe fires and their effects partly through drawings of a simple campfire. Simple examples help the reader unravel the complexity and apparent mystery of wildland fire.

Like you, I have been learning about fire since childhood. As an adult, I have studied and worked with fire, and its effects on plants and animals, for almost thirty years. I have taught people of every age about fire behavior and effects. Every time I want to explain fire better, I return to *The Book of Fire* for descriptions, definitions, and illustrations. When reporters and journalists ask about fire behavior, I show them drawings from this book. Until it went out of print about four years ago, I also suggested they obtain a copy. Luckily for us, the author and Mountain Press have made the book available once again, with revisions to increase its clarity and completeness.

As my understanding of wildland fire has increased, my curiosity and respect for its amazing power have also increased; I hope *The Book of Fire* brings you the same experience.

JANE KAPLER SMITH, Ecologist
Fire Sciences Laboratory, Rocky Mountain Research Station
U.S. Department of Agriculture, Forest Service
Missoula, Montana
September 2003

Preface

If it can't be expressed in figures, it is not a science; it is opinion.
—Robert A. Heinlein

If it can't be explained with pictures, it can't be explained to me.
—William H. Cottrell Jr., M.D.

Since Prehistory, people have stared into fire and wondered about flames, glowing coals, white ash, and charcoal. Why does a fire look like it does? How does fire work? This book tries to provide understandable answers to many of the tantalizing questions about fire. Though fire is deceptively simple in appearance, understanding it requires that we understand the elemental processes by which fire transforms a candle, campfire, or forest into gases, charcoal, and ash. Visualization of the unseen universe of atoms and molecules helps us immensely with this understanding.

This book provides the reader with scientifically accurate illustrations and explanations about fire. A glossary provides definitions for fire terminology.

Part I introduces the fundamental concepts of molecular physics, chemistry, and biology necessary to understand the flow of energy from its origin in the sun to its capture and storage by living cells and its subsequent release by the chemical reaction known as *fire*.

Part II visually describes the combustion process, exploring the anatomy and physiology of fire.

Part III applies concepts learned in Parts I and II to the real world of fire in wildlands.

Part IV explores the aftermath of wildland fire and the clues that reveal much about the fuels burned and the manner in which they burned.

PART I
Atoms, Molecules, and Chemical Reactions

The earth was created a living creature endowed with a soul and intelligence by the providence of God.
—*Plato*

The release of energy from fuel takes many forms. Forest fires and internal combustion engines extract their energy from fuels rapidly. By comparison, the living cells of plants and animals derive their energy from food more slowly. Irrespective of the rate of energy release, plants, animals, engines, and fire all use similar chemical reactions.

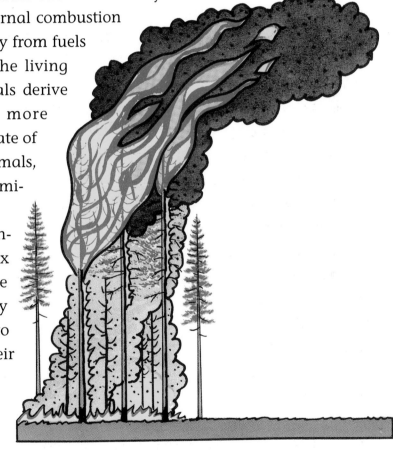

The release of stored energy by fire involves complex chemical reactions, so the appreciation of fire's mystery requires an introduction to atoms, molecules, and their interactions.

Earth, water, and air represent the three physical states of matter: solid, liquid, and gas. Nature uses tiny bits of matter called *atoms* and *molecules* to make mountains, forests, rivers, and air.

The earth and its orbiting moon resemble the nucleus and electrons of an atom. The earth is larger than the moon, and an atom's nucleus is enormous compared to its electrons. The moon orbits the earth at a great distance, just as electrons may travel far away from an atom's nucleus. Compared to electrons, the moon moves in a slow and predictable orbit; electrons whiz about a nucleus in an unknown fashion. They travel so astonishingly fast that they appear to be everywhere at once, forming a spherical, cloudlike shell around the nucleus.

orbit

Earth

Moon

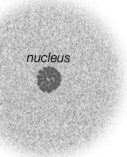

nucleus

electron cloud

Atoms combine with other atoms to form molecules by merging their electron clouds and sharing their electrons. Even the largest molecules are exceedingly small, measuring one hundred millionth of an inch in diameter and weighing one million trillionths of a pound. A teaspoon (one cubic inch) of air contains four hundred thousand trillion molecules of various gases, mostly nitrogen and oxygen.

One *atom* of oxygen and two *atoms* of hydrogen combine to form one *molecule* of water.

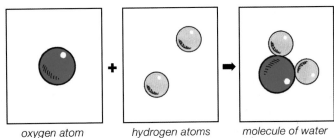

oxygen atom   hydrogen atoms   molecule of water

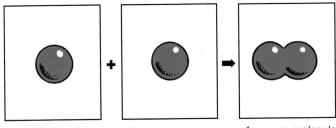

2 oxygen atoms   1 oxygen molecule

Oxygen from the atmosphere usually consists of two atoms combined to form one molecule of oxygen.

Six atoms of carbon, six atoms of oxygen, and twelve atoms of hydrogen combine to form one molecule of sugar.

LEGEND

= hydrogen atom

= oxygen atom

= carbon atom

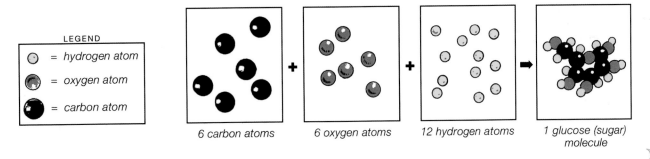

6 carbon atoms   6 oxygen atoms   12 hydrogen atoms   1 glucose (sugar) molecule

Unless "frozen" into a solid state of matter, molecules bounce about, randomly colliding with other molecules. Air's molecules collide with each other about five billion times per second. Heat energy agitates the molecules into continuous activity. Without the energy of heat, molecular motion ceases.

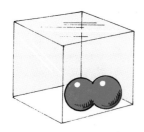

This single molecule of oxygen (two atoms) in a container has been cooled to absolute zero degrees Kelvin (-273°C or -460°F). No heat energy transfers to the molecule and it lies motionless.

If heat is applied to the container, heat energy transfers to the molecule, which starts to vibrate and bounce about the container. The molecular motion increases with continued heating.

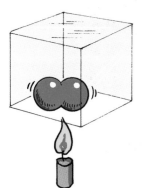

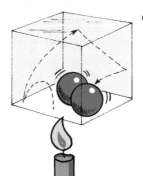

If the container held six molecules, they would collide with each other. The number of collisions per second (rate) would be determined by the amount of heat absorbed by each molecule.

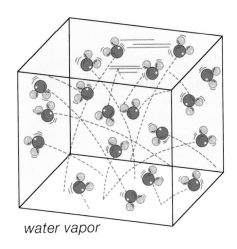

water vapor

Water molecules provide an excellent example of molecular behavior. Water molecules in this container are bouncing randomly in their gaseous state, which is called *water vapor*. Water molecules possess a lot of energy in their vapor state. It takes much more energy for a molecule to "fly" than to "swim."

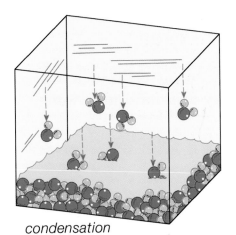

condensation

When water vapor molecules cool by giving off some of their heat energy, they slow down and undergo a process called *condensation*. The cooler, less energetic molecules collect as a liquid, such as clouds or dewdrops. In the liquid state, they still move about, but not nearly as much as when in the gas/vapor state.

ice

With further heat loss (cooling), the molecules slow even more and freeze into the solid form of water molecules called *ice*. Freezing traps the molecules in a network of solid crystals and restrains most of their movement.

If the frozen-solid water molecules are now warmed, each one will start to vibrate as it absorbs heat energy.

When sufficiently heat-energized, the molecules transform from the solid crystal state to liquid water.

Continued heating transforms the liquid water into water vapor. The rapidly moving water vapor molecules exert enough pressure to lift the pot's lid and escape into the air.

The term *chemical reaction* describes the changes that occur when different types of atoms and molecules interact.

All chemical reactions require energy from an outside source. Energy stirs the molecules and initiates the reaction. For example, frozen oxygen doesn't combine with frozen hydrogen because the molecules have too little energy for motion or mixing.

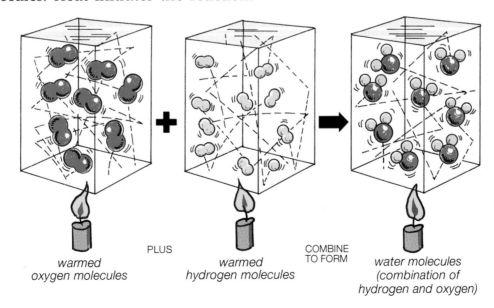

frozen oxygen molecules

frozen hydrogen molecules

MIXED WITH

no reaction

Warm molecules vibrate and bounce about freely, so they mix well. They also interact readily in chemical reactions if appropriate energy conditions exist. A mixture of gaseous oxygen and hydrogen molecules, if heated with a flame, reacts rapidly. The molecules exchange electrons to form water molecules. Heat initiates the reaction.

PLUS

COMBINE TO FORM

warmed oxygen molecules

warmed hydrogen molecules

water molecules (combination of hydrogen and oxygen)

ENERGY

Ultimately the sun provides the energy required to keep molecules and atoms interacting in chemical reactions.

Within the sun's fiery interior, unimaginably intense heat and immense pressure fuse four atoms of hydrogen. This fusion of four atoms creates one atom of helium. The fusion process (like the hydrogen bomb) releases tremendous quantities of energy as heat and light. Heat and light then travel through outer space to earth.

Energy Transfer

To be useful, energy must be transferred from point A to point B. Molecule A (George) has three options for transferring energy (boiling-hot potato) to molecule B (Sandy). He may hand it to others to pass it on (conduction), lob it directly through the air (radiation), or run over and hand-deliver it himself (convection).

Molecules have the same three choices when transferring energy.

Conduction: Molecule A "passes" its energy to other molecules, which transfer it eventually to B. For example, heat travels through a metal rod by conduction.

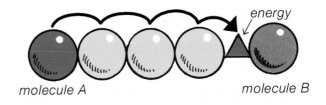

molecule A   molecule B

Radiation: Molecule A sends its energy through space to B. The sun transfers its energy to the earth by radiation.

molecule A   molecule B

Convection: Energized molecule A "carries" its energy directly to molecule B.

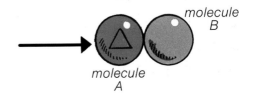

molecule A   molecule B

The skier warms her hands with heat transferred by radiation. The stove heats the kettle of water by conduction. Hot gas molecules moving up the stovepipe heat it by convection. The pipe then radiates its heat.

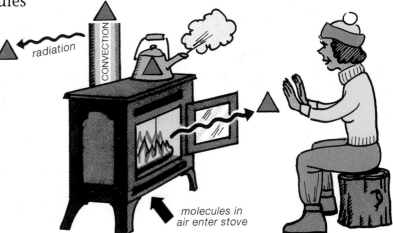

radiation

CONVECTION

molecules in air enter stove

THE PRODUCER REACTION
(Photosynthesis)

To make food, green plants collect energy from the sun through leaves, needles, and sometimes twigs, trunks, and roots. Food is a useable and storable form of chemical energy. Plants absorb carbon dioxide from the air, and with their roots, absorb water from the ground. The plants join the sun's energy, carbon dioxide, and water in a chemical reaction called *photosynthesis*. Photosynthesis, which means "made with light," is also called the *producer reaction*.

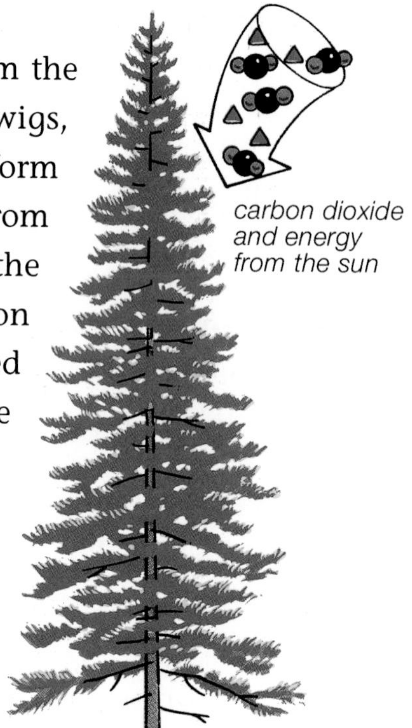

carbon dioxide and energy from the sun

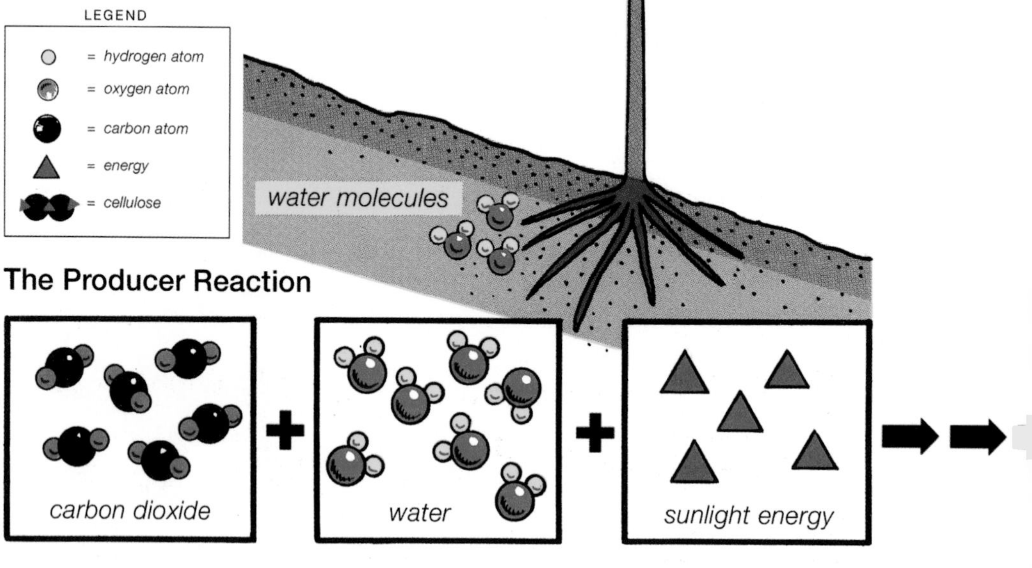

LEGEND

◯ = hydrogen atom

◉ = oxygen atom

● = carbon atom

▲ = energy

= cellulose

water molecules

The Producer Reaction

carbon dioxide + water + sunlight energy ➡ ➡

Photosynthesis assembles carbon atoms into sugar molecules. Sugar is then used in products of varying size and shape, like starch, fat, and cellulose. A bit of the sunlight energy is captured and held each time one carbon atom bonds to another carbon or hydrogen atom. Plant components like sugar, fat, and cellulose contain many such bonds. Photosynthesis also stores some of the light's energy in oxygen, which the plant releases into the air.

oxygen released to air

pine needle

water + carbon dioxide + energy

SUGAR, STARCH, FAT, CELLULOSE, RESIN, ETC.

cellulose

sugar molecule

energy from sun

Cellulose: *a long stringy molecule composed of 20,000 sugar molecules (more than 100,000 carbon atoms).*

Sugar molecule: *the basic building block for many plant products like starch and cellulose.*

Energy from the sun is captured between bonded carbon atoms, and in bonds between carbon and hydrogen.

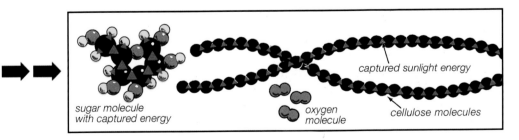

sugar molecule with captured energy

oxygen molecule

captured sunlight energy

cellulose molecules

THE CONSUMER REACTION
(Metabolism and Combustion)

To obtain energy, animals must eat plants that have made and stored lots of food; therefore, animals are called *consumers*. Consumers employ a chemical reaction called *metabolism* to release the energy stored in plants. Metabolism disassembles the products of photosynthesis and releases the energy stored between carbon atoms, and between carbon and hydrogen atoms.

The bison's stomach digests the cellulose of grass into small sugar molecules. Sugar next travels to every cell of the bison's body. There, sugar combines with inhaled oxygen to release the energy stored in sugar's chemical bonds. The bison uses the released sunlight energy for all of its activities. Finally, it exhales carbon dioxide and water.

cellulose

carbon dioxide and energy

oxygen

oxygen

carbon dioxide

The Consumer Reaction

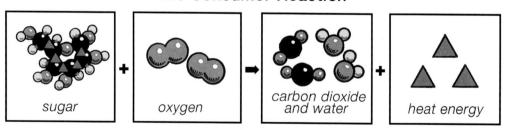

sugar + oxygen → carbon dioxide and water + heat energy

These campers roasting marshmallows demonstrate the consumer reaction carried out at different rates—slow and fast.

The eager camper on the left will soon devour his marshmallow. Then he will carry out the slow consumer reaction and metabolize the marshmallow's sugar into carbon dioxide and water. The sunlight energy stored in the sugar molecules will be slowly released.

The disappointed camper on the right watches the energy stored in his marshmallow sugar being released all at once by fire—a rapid consumer reaction called *combustion.*

Metabolism = slow consumer reaction
Combustion = rapid consumer reaction

A bison inhales oxygen for use in its metabolism. Metabolism releases the stored energy in inhaled oxygen molecules and in food molecules. Energy released from those molecules heats the bison's shivering body. The bison exhales a moist cloud of carbon dioxide and water.

Some consumers employ a chemical reaction called combustion to rapidly release energy stored in plants. For example, the snowmobile needs "food" just like a bison. The machine "inhales" oxygen to mix with carbon-based fuels like oil and gasoline (long-dead producers). Combustion in the snowmobile's engine cylinders splits the "inhaled" oxygen molecules and the fuel's carbon molecules— releasing the energy stored in both.

The bison's metabolism of food and the snowmobile's combustion of fuel are similar chemical reactions. Each uses oxygen to efficiently release the stored energy made by plant producers.

FIRE—A CHEMICAL REACTION

Fire is a rapid and persistent chemical reaction that combines fuel and oxygen to produce heat and light. An external source of heat called the pilot heat is usually needed to start the reaction.

In a snowmobile engine, gasoline fuel and oxygen from the air require the heat of an electrical spark to ignite the fuel-oxygen mixture and start the engine.

The Fire Reaction

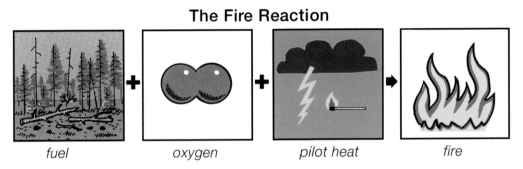

fuel   oxygen   pilot heat   fire

In forests and prairies, plants (living or dead) and oxygen from the air also require pilot heat to ignite them. This pilot heat can be supplied by lightning, volcanoes, or humans. Like animal metabolism, the fire reaction requires both fuel and oxygen for its continued existence. If the fire runs out of fuel or oxygen, it goes out. Both fuel and oxygen are needed to release the energy that is stored in their chemical bonds.

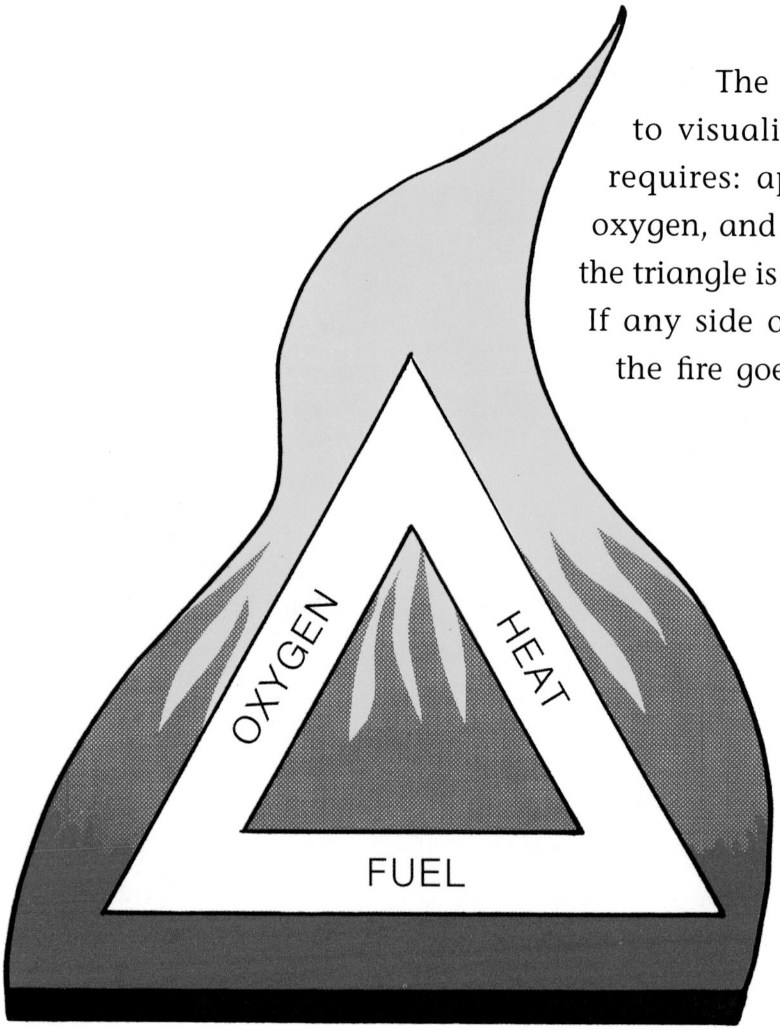

The fire triangle is a good way to visualize the three things a fire requires: appropriate fuel, adequate oxygen, and enough heat. If any side of the triangle is missing, a fire cannot start. If any side of the triangle is removed, the fire goes out.

OXYGEN

HEAT

FUEL

PART I *Atoms, Molecules, and Chemical Reactions*

PART II
Fire, Flame, and Combustion

Of design, the earth was created to provide its own food from its own waste, and all that it did or suffered was turned back on itself.

—Plato

Fire and flame do not share the same meaning. Fire is the chemical reaction of combustion. There are two main types of combustion: flaming and glowing/smoldering. Flames are visible only during flaming combustion, when wood-fuel gases mix with oxygen in the air and burn above the original fuel surface. Nonflaming combustion is divided into two types, glowing and smoldering. Glowing embers display little or no flame, although combustion is occurring. In smoldering combustion, fuels like peat, duff, roots, and stumps burn without flame or embers. Glowing and smoldering combustion are supported by oxygen that travels to the surface of wood-fuel solids and densely packed fuels. Flaming consumes fuel gases, and glowing/smoldering consumes fuel solids.

A burning tree demonstrates the transition from flaming to glowing combustion.

FIRE

When wood burns, it passes through two very distinct phases. A visually athletic flaming phase consumes the potential gases stored in the wood. A prolonged and subtle glowing/smoldering phase releases the energy stored in the solid fuels after flaming subsides.

During flaming combustion, the woody fuel emits gases that mix with air above the fuel. The chemical reaction of flaming combustion consumes the mixture of fuel gases and air.

When the flame-heated fuel runs out of gases, the chemical reaction descends to the fuel's surface.

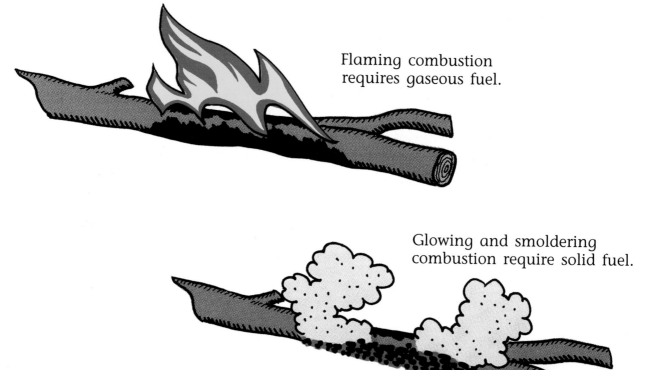

Flaming combustion requires gaseous fuel.

Glowing and smoldering combustion require solid fuel.

FLAME

Flame may be visualized as a bag filled with hot fuel gases. Oxygen surrounds the bag's exterior. The walls of the bag provide a meeting ground for the interaction of fuel and oxygen.

The flexible bag of flame contains unburned fuel gases and soot (partly burned particles of fuel). Within the thin bluish walls of the flame bag, fuel and oxygen combine. This chemical reaction splits the bonds in the fuel and oxygen to release the energy originally acquired from the sun.

Some of that energy heats the soot inside the flame bag, causing it to radiate heat and light, much like a hot lightbulb element.

The wall of the flame bag is called the *combustion reaction zone*. If this reaction zone fails, the flame smokes and dies. Complex chemical reactions progress within the body and wall of the flame bag.

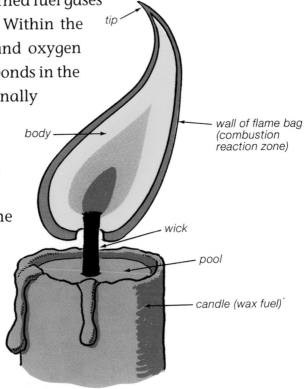

tip

wall of flame bag (combustion reaction zone)

body

wick

pool

candle (wax fuel)

A burning candle illustrates the structure of flame and the process of flaming combustion.

① Fuel (liquid wax) migrates up the wick from a pool at the top of the candle. It then boils off the wick. Transformed into a fuel gas, it "inflates" the fuel bag and circulates before entering the combustion reaction zone.

② Fuel gas molecules enter the fuel bag walls (the reaction zone) and mix with oxygen from the surrounding air. Chemical bonds in the fuel gas and oxygen split, thereby releasing their energy.

③ Carbon dioxide, soot, and water produced in the reaction zone float away.

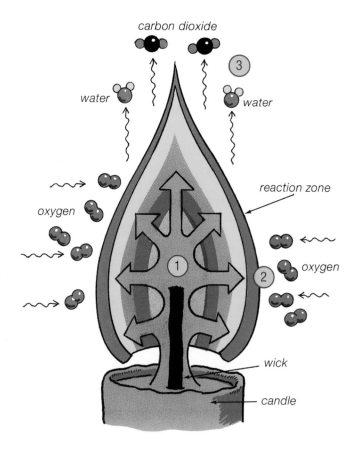

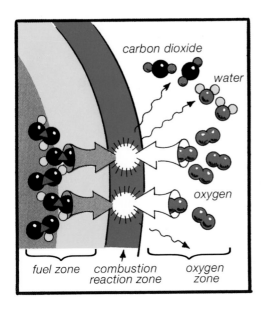

carbon dioxide

water

oxygen

fuel zone   combustion   oxygen
reaction zone   zone

This diagram shows the combustion reaction zone positioned between the fuel gases and oxygen. Combustion occurs when fuel gases combine with oxygen in the reaction zone.

Burning a ball of crumpled paper further demonstrates the principle of the flexible flame bag. Fuel gases pour from numerous chambers and folds of the crumpled paper and inflate the bag into many fantastic shapes. The flame floats on the invisible fuel gas molecules. When the paper no longer produces fuel gas, the flame bag will collapse onto the paper. The flame bag's reaction zone will spread across the paper and burn its solid fuel by glowing combustion.

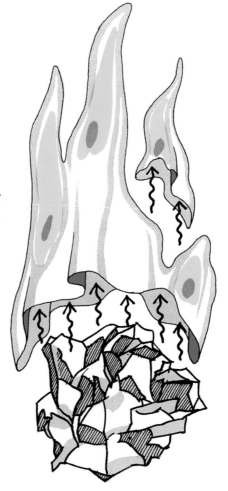

PYROLYSIS PATH AND SOOT CYCLE

Fuel Behavior inside a Flame

(1) Fuel vaporizes out of the wick and enters the flame's interior as long chains of carbon atoms. The fuel molecules may randomly travel the pyrolysis path or enter the soot cycle.

Pyrolysis Path

(2) *Pyro-lysis* means "heat-divided." The intense heat radiating from the reaction zone pyrolyzes the fuel. That means the fuel chains are subdivided into smaller and smaller fragments. Pyrolysis of fuel makes it easier to burn, like kindling split from a large round of wood.

Pyrolysis reduces the large fuel molecules to fragments 2 to 4 carbons long. Then they enter the reaction zone, where they mix with oxygen molecules. When the chemical bonds in the fuel and oxygen break, they release heat energy originally obtained from the sun. The by-products, carbon dioxide and water, migrate away from the reaction zone.

(3) A delicate blue flame accompanies the combustion of fuel gas in the reaction zone.

Soot Cycle

(4) Some unburned carbon atoms in the flame bag cluster into snowball-like particles called *soot*. These conglomerations of fuel fragments and other chemical substances from the brew circulating in the flame's interior disperse throughout the flame bag.

Soot particles, like lightbulb filaments, radiate light of varying colors and intensities, depending on the amount of energy they absorb. Soot within the flame bag absorbs energy from the reaction zone and radiates a yellowish to orange light, depending on its proximity to the reaction zone. Soot near the wick absorbs little heat, radiates little or no light, and appears black.

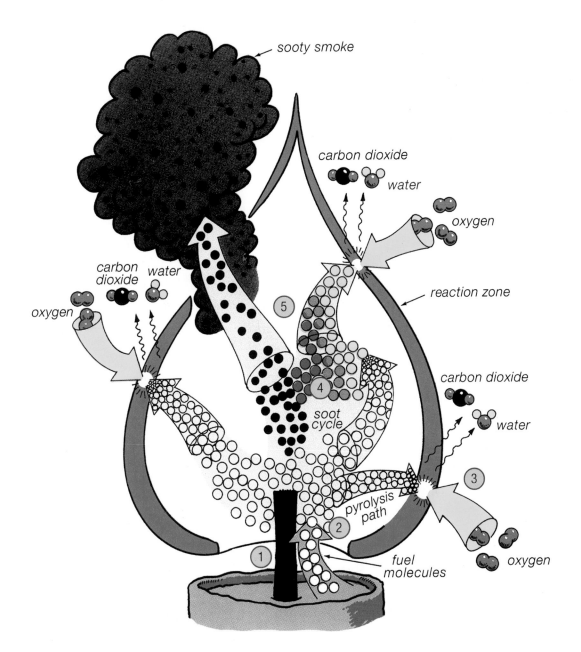

(5) The soot particles may either enter the reaction zone and be combusted, producing heat, carbon dioxide, and water, or they may exit the flame bag through an inefficient or absent section of reaction zone and immediately cool to black, sooty smoke.

FLAME IGNITION

Pilot heat is needed to start a fire. In wildlands, pilot heat originates from some external source like lightning, volcanoes, chainsaw sparks, neglected campfires, cigarettes, and matches.

Pilot-Dependent Fire Reaction:

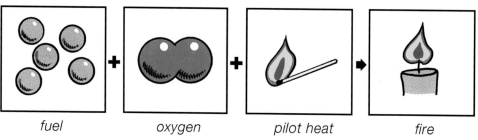

fuel　　　　oxygen　　　　pilot heat　　　　fire

Initiation of flaming combustion requires gaseous fuel. In the example of a burning candle, the match's flame (pilot heat) melts and vaporizes the wax to form a cloud of fuel gas (flame bag) around the wick.

Molecules of oxygen and fuel gas react most rapidly near the pilot flame because its heat increases the reaction speed.

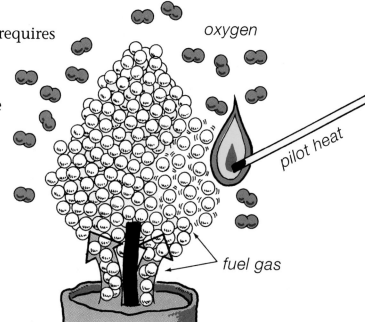

oxygen

pilot heat

fuel gas

Eventually, enough fuel molecules split, bond with oxygen, and release their stored energy to initiate a continuous reaction. Molecules start exploding apart like a book of burning matches. The bluish combustion reaction zone suddenly bursts into life. The zone spreads rapidly in all directions from the pilot heat. The zone encloses the cloud of fuel gas to form the flame bag.

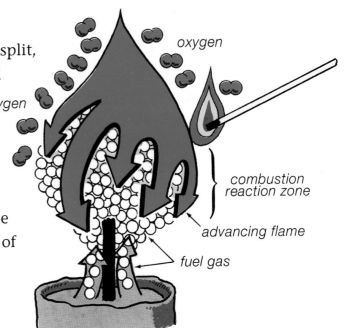

oxygen

oxygen

combustion reaction zone

advancing flame

fuel gas

The reaction zone soon produces enough heat to sustain combustion without the match's pilot heat. The fire burns independent of the pilot heat and produces its own heat.

Independent Fire Reaction

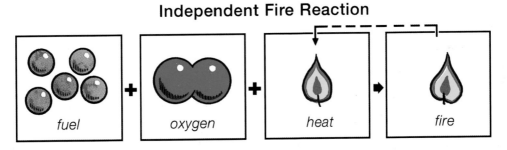

fuel + oxygen + heat → fire

Flaming combustion requires gaseous fuel, which travels to the combustion reaction zone. *Solid fuel burns in the opposite manner;* the reaction zone must travel to the fuel.

PYROLYSIS

Wildland fuels burn less efficiently than wax candles. For example, efficient pyrolysis of wax mainly yields gaseous molecules that are small enough for flaming. Ineffecient pyrolysis of wildland fuels (for example, wood, leaves, and pine needles) produces two products: (1) gaseous molecules that either burn by flaming combustion or accumulate as a smoke of small black droplets called *tar*, and (2) blackened solids called *char* that coat the surface of underlying unburned fuel. This inefficient pyrolysis is similar to scorching, a term familiar to people who have burned their toast, scorched their shirt, or smoked a cigarette.

Defective toasters deliver charred bread in a cloud of tarry smoke. The bread did not burn but scorched, due to pyrolysis.

A hot iron may alter the cellulose structure of cotton fabrics, resulting in a permanently charred surface and a puff of tarry smoke. The fabric scorched but did not burn.

The cigarette smoker inhales the tarry smoke of scorched tobacco cellulose.

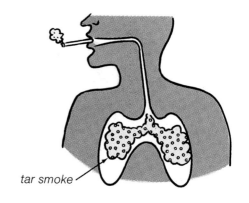

Scorching of wildland fuels produces tar droplets and soot particles that enter the gaseous fuel cloud and form smoke.

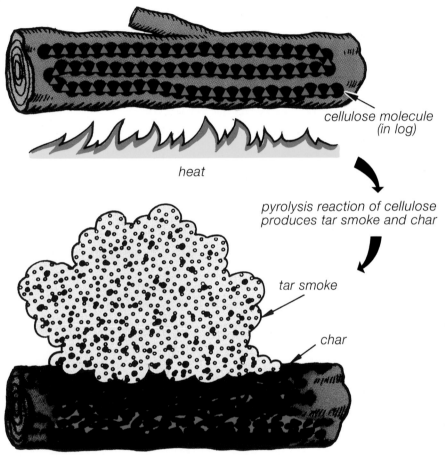

cellulose molecule
(in log)

heat

pyrolysis reaction of cellulose
produces tar smoke and char

tar smoke

char

Char forms a black layer of carbon atoms on the heated cellulose surface. Charred fuel and tar smoke are both fuels with many high-energy bonds remaining between their carbons and hydrogens. With continued heating, tar smoke may ignite in flame and char surfaces may begin to glow.

PHASES OF COMBUSTION

Wildland fuels pass through three phases of combustion: pre-ignition, ignition, and combustion. Many fuels initially burn by flaming combustion (burning of fuel gases), followed by glowing and smoldering combustion in solid fuels.

Solid fuel combustion is best appreciated by observing the manner in which a piece of wood burns.

Pre-Ignition Phase

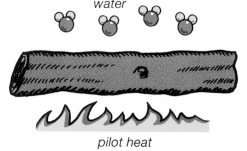

water

pilot heat

gas molecules   fuel cloud

pilot heat

The pilot heat first dries the fuel by boiling off moisture contained within it. If the pilot heat is not adequate to dry the fuel, combustion stops. Rain-soaked wood has a prolonged pre-ignition phase, while drought-dried wood can have a dramatically brief pre-ignition phase.

With continued heating, the fuel absorbs more energy and boils off substances in the wood like fat, oil, terpene, alcohol, and resin. These collectively form a cloud of fuel molecules.

Persistent application of heat scorches the fuel, producing a blackened surface of char that emits a gray smoke of tar.

Tars occur in gaseous form and in condensed droplets. These contribute greatly to the cloud of fuel gas. At this point the wood has scorched, not burned.

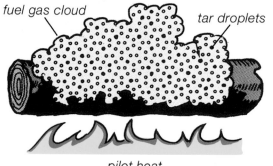

fuel gas cloud   tar droplets

pilot heat

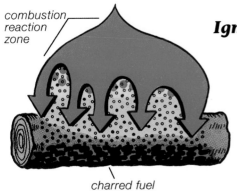

combustion
reaction
zone

charred fuel

Ignition Phase

The pilot heat ignites the fuel cloud. The combustion reaction zone materializes and rapidly encloses the fuel gas cloud. The fire now burns independent of the pilot heat. The combustion reaction is self-sustaining.

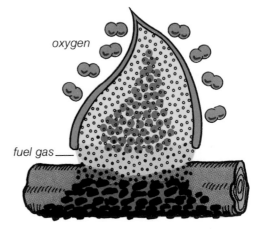

oxygen

fuel gas

Combustion Phase

The radiant heat of flaming combustion continues to release gas molecules from the fuel source. These gases rise up to enter the fuel cloud and eventually burn. Radiant heat also pre-ignites adjacent unburned wood.

After fuel molecules within the surface of the wood have pyrolyzed to fuel gas, char, and tar, the production of fuel gas slows. The flame bag dwindles and collapses like an improperly pitched tent.

The reaction zone *must travel to the fuel* in order for combustion to continue.

The collapsed reaction zone spreads out on the charred wood surface and initiates glowing/smoldering combustion.

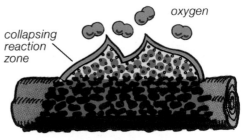

oxygen

collapsing
reaction
zone

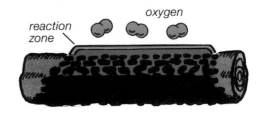

oxygen

reaction
zone

When the combustion reaction zone settles onto the charred surface of the wood fuel, it spreads like a heavy mist into the black network of carbon molecules. The reaction zone continues to provide opportunity for the charred fuel and atmospheric oxygen to join and burn.

The reaction zone heats the char to reddish orange glowing embers. This is the beginning of *glowing combustion*. The vibrating molecules disrupt their connecting bonds and combine with oxygen. As they combust, they release their stored energy along with water vapor and carbon dioxide.

The heat of glowing combustion pyrolyzes adjacent unburned wood to produce fuel gases and char, which will sustain the process of glowing combustion. When the heat of glowing fails to ignite the fuel gases, they condense into large volumes of tarry smoke.

After the reaction zone burns away all of the fuel, only a delicate network of whitish mineral ash remains. The ingredients of ash resemble those of sand and do not burn.

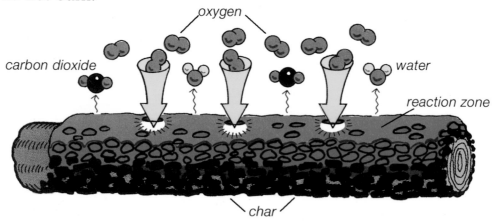

Mortar has little strength after the bricks are removed. Likewise, after carbon atoms leave the wood, ash has no strength and collapses. Ash may form a layer on unburned fuel that blocks oxygen from reaching the reaction zone.

The reaction zone subsides, as does the glowing combustion under too thick a layer of ash.

Glowing combustion usually progresses from the bottom to the top of tree branches and trunks lying prostrate on the ground. This is because the mineral ash falls away from the reaction zone, thereby giving oxygen easy access to the unburned fuel. Stirring and repositioning wood in a campfire knocks away the smothering ash and may even renew flaming combustion.

Sometimes fire produces enough heat to dry and pyrolyze the fuel (pre-ignition), but something keeps the fuel from burning completely. Perhaps the fire is running low on fuel. Perhaps the fuel is so tightly packed that oxygen is slow to reach its surface. Maybe a layer of white ash is keeping most oxygen out. The fuel may be so moist that most of the fire's heat is spent drying it out. In any case, the pyrolyzed fuel gases condense into tarry smoke, but no flame or glowing occurs. This process is called *smoldering combustion.*

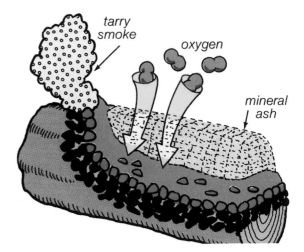

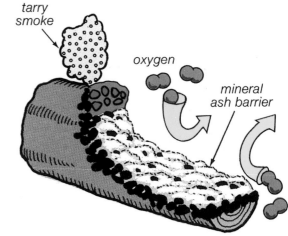

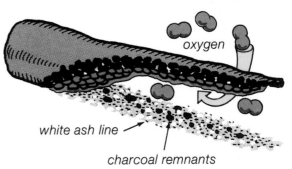

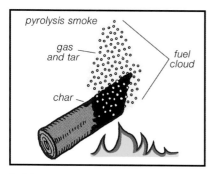

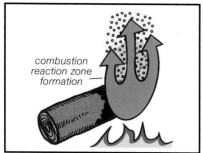

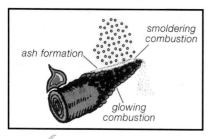

Phases of Combustion—Summary

Pre-ignition phase—dehydration and fuel cloud formation. Pilot heat first dries the fuel. The heat next pyrolyzes the fuel, forming a cloud of flammable gases and tar droplets. The fuel surface chars.

Ignition phase. The combustion reaction zone forms and envelopes the fuel cloud. Combustion is now self-sustaining, and the pilot heat is unnecessary.

Combustion phase—flaming and glowing/ smoldering. Flaming combustion consumes the fuel gases produced from wood, leaves, and other fuels. The flames preheat the adjacent unburned fuel, which emits more gaseous fuel. The fire spreads when this fuel cloud ignites. Flames dominate this phase of combustion.

Glowing and smoldering combustion dominate when the reaction zone settles onto the fuel surface. Charred fuel combines with oxygen and burns as glowing embers. Even when heat or oxygen is insufficient to burn the fuels completely, pyrolysis may continue, producing char and tarry smoke.

PART III
Fire in Forests

*Unless one comes to an understanding concerning the
nature of change, one will have many difficulties.*
—Plato

Fuel, terrain, and weather determine the type and behavior of individual wildland fires. The principles of chemical reactions and combustion learned in Parts I and II are important to understanding how a leaf, twig, tree, or forest burns. The focus is on fire in forests because their structure and fuel variety provide the most complex and powerful fires observed in wildlands. But the principles of fire spread described in this chapter apply just as well to other fuels, such as those occurring in prairies, shrublands, or a backyard's pile of leaves.

In this chapter, illustrations show the layers of fuel in a forest. The location and character of these fuels help determine if a fire will burn in the treetops or on the forest floor, and what kind of smoke will be produced.

FOREST STRUCTURE

A wildland forest consists of a plant community dominated by trees growing on the forest floor. The tops, or crowns, of trees form the top layer or canopy, which covers part or most of the ground. Small plants, such as grasses, wildflowers, and shrubs, comprise the forest's understory.

The forest floor consists of litter, duff, and mineral soil. The top layer, called litter, contains dead needles, leaves, and twigs that don't yet show signs of decomposition. The duff layer is composed of recent and past years' dead and decomposing plants and animals. Mineral soil develops from decomposition of underlying rocks like granite, limestone, sandstone, or basalt.

The duff and upper mineral soil contain roots, seeds, bacteria, and fungi. Burrowing worms and moles mix the layers.

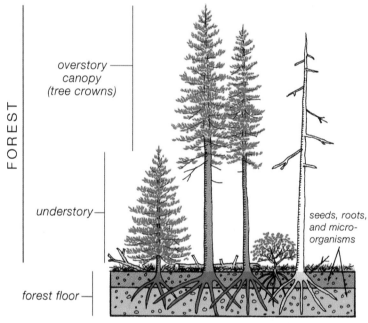

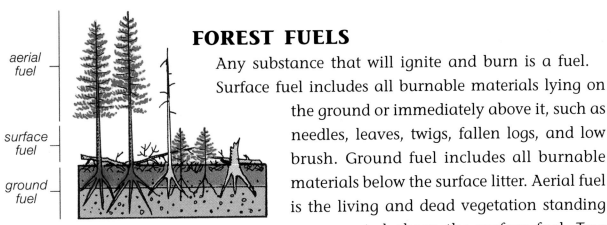

aerial
fuel

surface
fuel

ground
fuel

FOREST FUELS

Any substance that will ignite and burn is a fuel. Surface fuel includes all burnable materials lying on the ground or immediately above it, such as needles, leaves, twigs, fallen logs, and low brush. Ground fuel includes all burnable materials below the surface litter. Aerial fuel is the living and dead vegetation standing or supported above the surface fuel. Tree crowns are an important part of aerial fuels.

Fuel Characteristics

Dead or living. More than half the weight of living fuels is water used by the plant for biological processes.

Dry or wet. Fuel must be dry to burn.

Fine or heavy. Fine fuels measure less than $\frac{1}{4}$ inch across. They dry rapidly, ignite easily, and carry fire to heavy fuels.

FIRE TYPES

We classify wildland fire behavior according to the fuel layers in which the fire is spreading. Most fires are complex blends of ground fire, surface fire, and crown fire.

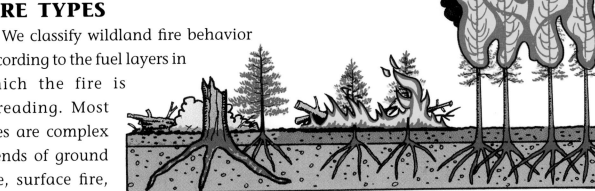

ground fire in ground fuel   surface fire in surface fuel   crown fire in aerial fuel

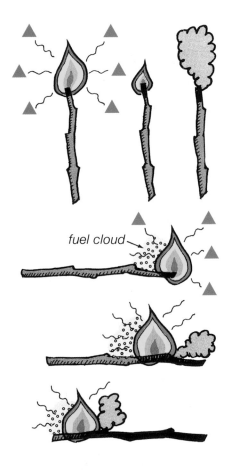

fuel cloud

FUEL POSITION AND FIRE BEHAVIOR

A burning twig illustrates how fuel position determines the spread of fire.

In the vertical position, a twig quickly fizzles out. Radiation is coming from the flame bag, which is mostly above the fuel, so only a little heat is radiating from the flame to unburned fuel. Most of the flame's radiation, and all of its convective heat, travel upward, away from the fuel. The portion of the twig below the flame is not heated enough to form fuel gases, so combustion stops.

In the horizontal position, more radiant and convective energy is available to preheat the unburned portion of the twig. The preheated wood generates a flammable fuel cloud and completes pre-ignition. The existing flame serves as pilot heat for the ignition phase in the new fuel cloud. The combustion phase follows quickly, and the fire spreads slowly but surely along the twig.

In the inclined position, the radiating flame body is much closer to the unburned wood, and the convection column from the flame actually surrounds some of the unburned fuel. The unburned wood is rapidly preheated, ignited, and burned.

The distance from the flame bag and convection column to the unburned fuel greatly affects the rate of fire spread. A short distance means rapid spread.

These principles apply readily to wildland settings. In a wildland fire, unburned fuels absorb heat from burning fuels. The transfer of heat from burning fuel to unburned fuel occurs by radiation and convection *(see pages 8–9)*.

The distance between flame and fuel determines the amount of energy absorbed from radiation. When the flame and fuel are close to each other, fuel absorbs much more energy from radiation.

Convective heat transfers energy to fuels above the fire. Hot molecules rise from the region of flame to the fuel and preheat it.

This example illustrates how a surface fire in fallen fuels spreads into aerial fuels. Flame 1 preheats a few trees at a time through radiation. Flame 2 preheats many more trees because the unburned fuels are above the fire, near the flame bag and convection column. The fire spreads rapidly, "running" up the slope. Flame 3 preheats little fuel by radiation and none by convection. Many wildland fires stop on sharp ridges and mountaintops.

fuel clouds

Wind pushes the flame bag and the convection column closer to unburned trees, preheating them rapidly. Uphill fire spread brings the flame bag and convection column closer still to the unburned trees. Wind encourages downhill fire spread by bending the flame bag close to fuels that would escape preheating if there were no wind.

fuel clouds

WIND

In this example, the fuels lie prostrate on the ground. Flame ① preheats very little fuel and may not ignite the adjacent scattered surface fuels. Flame ② preheats a lot of fuel through both radiation and convection. It will progress rapidly through the three phases required for combustion.

HOW A TWIG WITH NEEDLES BURNS

Pre-ignition phase—dehydration and fuel cloud formation

A pilot heat source dries the fuel by boiling off stored water. Then a fuel cloud develops, containing molecules derived from hot sap, resin, terpene, oils, and tar droplets. The twig and needles scorch, producing tar and char.

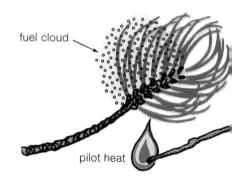

fuel cloud

pilot heat

Ignition phase

The pilot heat ignites the flammable fuel cloud. A combustion reaction zone spreads immediately from the ignition point. It envelopes the entire fuel cloud and establishes self-sustaining combustion.

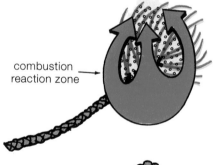

combustion reaction zone

Combustion phase—flaming

Flaming combustion rapidly depletes the tiny bag of fuel gases released by the twig and needles.

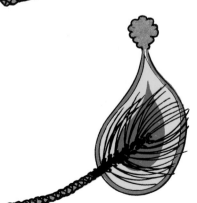

Combustion phase—glowing and smoldering

Glowing combustion consumes the solid fuel, which consists of charred needles and twigs.

Glowing combustion slowly converts most of the remaining char to carbon dioxide, water, and energy. White mineral ash is all that remains of the needles. If ash smothers glowing combustion in the twig, then the twig's incompletely burned and charred surface emits a small cloud of tarry smoke and goes out.

The fragile skeleton of mineral ash awaits disturbance.

The delicate ash debris drifts to the ground under gentle conditions or blows away with the wind. Strong wind may carry the ash hundreds of miles from the fire.

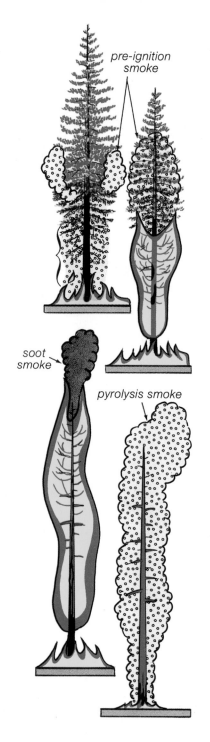

pre-ignition smoke

soot smoke

pyrolysis smoke

HOW A TREE BURNS

Torching describes the astonishing process of nearly instantaneous combustion of a tree. Very hot, dry weather promotes torching. Torching occurs in isolated trees or small groves of trees. When the intensely hot combustion reaction zone touches concentrated fine fuels like the needles in a pine tree's crown, the phases of combustion occur rapidly.

Burning surface fuel preheats the lower tree. The pre-ignition phase terminates with a hot fuel cloud, scorched bark and needles, and tarry smoke.

The ignition phase explosively consumes the fuel cloud and envelops the lower tree with flames. The upper tree is instantly preheated and enters a pre-ignition phase.

Flaming combustion envelops the entire tree and its fuel gas cloud. The flaming fuel cloud scorches the surface of the tree trunk, turning it black.

Glowing combustion persists in the charred surfaces left by flaming combustion, but it rapidly dies out in living trees with high water content. Smoldering may continue in pockets of rotten wood that ignited when the tree burned or in densely packed fine fuels in the duff surrounding the base of the tree.

Wind Effect and Two-tones

When surface winds blow strongly, the fuel cloud, caused by surface fire preheating, concentrates on the downwind side of the tree.

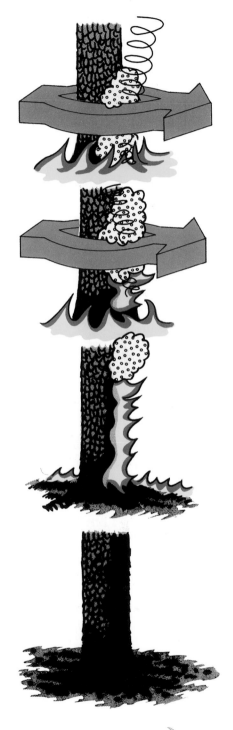

The concentrated fuel cloud ignites.

The flaming fuel cloud climbs the tree trunk and scorches its bark. As with a burning match, the flame preheats the tree immediately above it. Preheating releases flammable gases that also ignite.

The two-tone trunks reveal which way the wind blew when the trees burned. Two-tone trees contain mixtures of dead brown needles, scorched black needles, and even living green needles spared by the blowing wind.

HOW A FOREST BURNS

Most forest fires start from the pilot heat supplied by lightning or by people's matches and sparks. Fires usually start in surface and ground fuels. Environmental conditions must be right for a fire to then "climb" into a tree crown or forest canopy. A plentiful supply of dry fuel, ladder fuels, and dry, warm air facilitates a fire's vertical spread.

Fire climbs into tree crowns by using these fuel ladders:

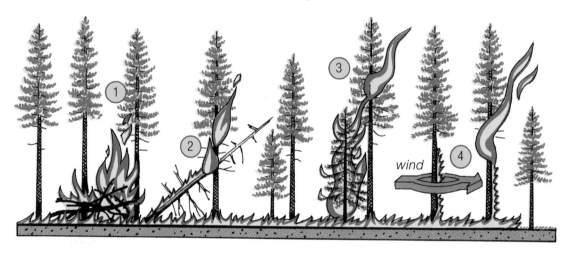

(1) Plentiful dry surface fuel burns hot enough to preheat and ignite aerial fuels in the crowns.

(2) Fire ascends partially fallen trees into the crowns.

(3) Surface fires ignite understory trees whose branches drape near to the ground, yet reach up to the forest canopy.

(4) If the living tree's water content is very low, as in a drought, and if there is a wind, a fuel cloud concentrates downwind of the trunk, ignites, and flames "climb" up into the tree's crown.

When a fire climbs into the tree crowns and many trees begin torching, the process is called *crowning.* If it spreads through the tree tops it is called a *crown fire.* Surface and crown fires together preheat unburned trees ahead of the moving flame front. This is called a *dependent crown fire.*

If the crown fire races ahead of the surface fire and if it no longer requires the surface fire's assistance to preheat fuel, then it is called an *independent crown fire.* This occurs in the presence of strong wind.

An independent crown fire can "make a run" of several miles in a single day.

Once a fire has started, it can spread by several means: crowning (see previous page), spotting, heading, or backing.

Sometimes wind transports burning aerial fuels far ahead of a main fire. When aerial fuels, called *firebrands*, fall to earth, they may ignite a spot fire.

A fire spreads from the point of ignition in three directions, which are called *fire fronts.* A heading fire front spreads rapidly in the direction of the wind or uphill. A backing fire front slowly creeps into the wind or downhill. A flanking fire front spreads at right angles to the wind or across a slope.

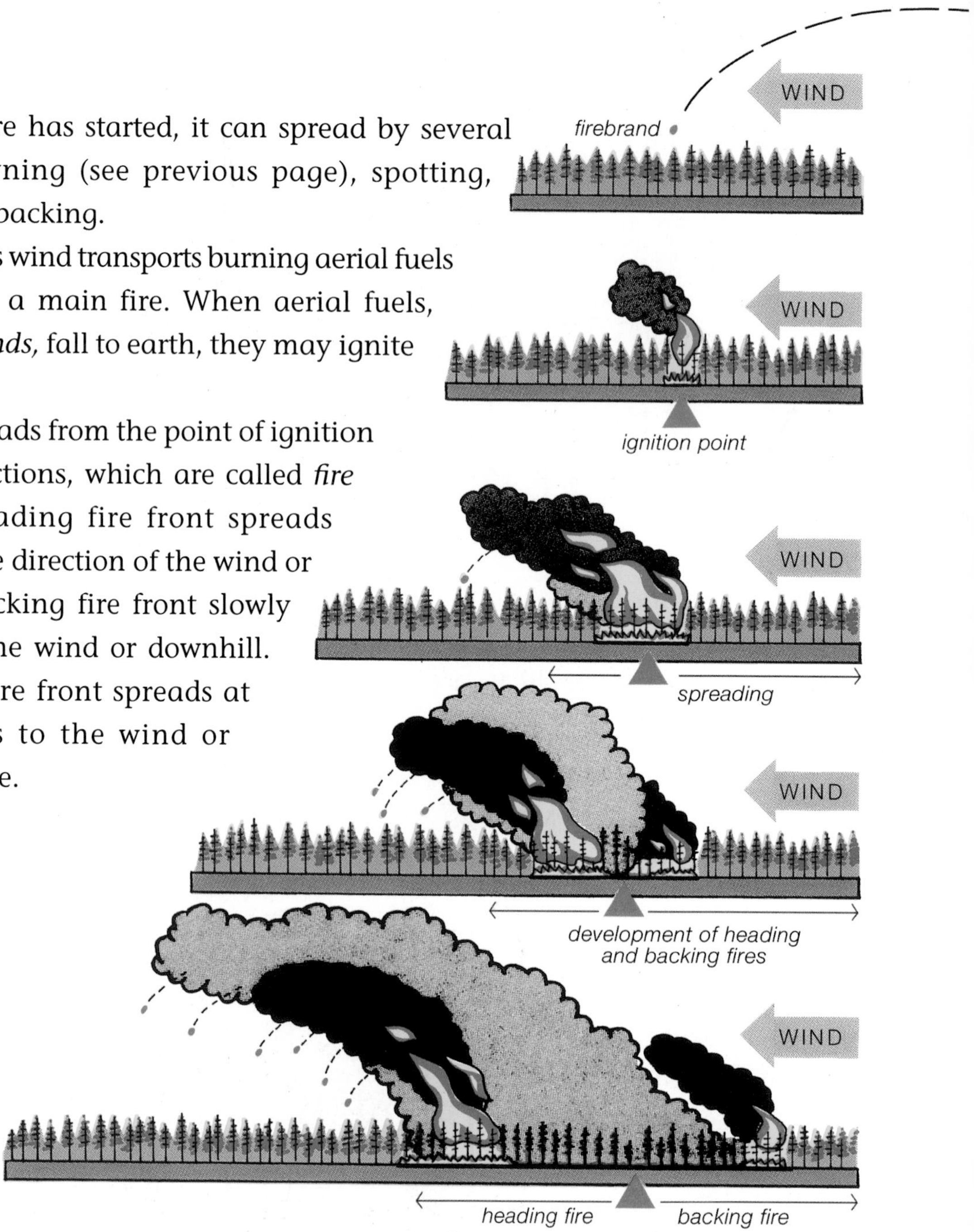

firebrand

WIND

ignition point

WIND

WIND

spreading

WIND

development of heading and backing fires

WIND

heading fire   *backing fire*

PART III *Fire in Forests*

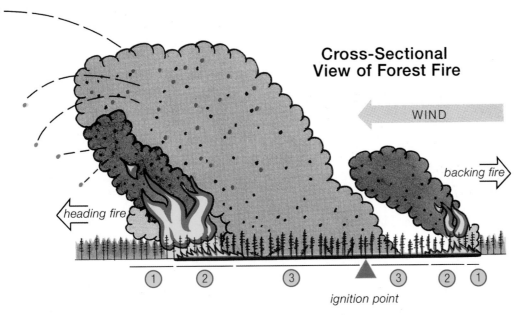

**Cross-Sectional
View of Forest Fire**

WIND

backing fire

heading fire

① ② ③ ▲ ③ ② ①

ignition point

These diagrams show a typical forest fire, with heading, backing, and flanking fires, their direction of movement, and their three phases of combustion: ① pre-ignition, ② ignition and flaming combustion, ③ glowing and smoldering combustion.

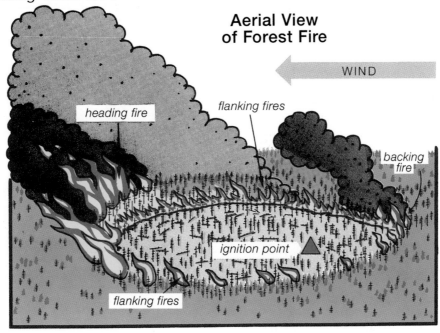

**Aerial View
of Forest Fire**

WIND

heading fire

flanking fires

backing fire

ignition point

flanking fires

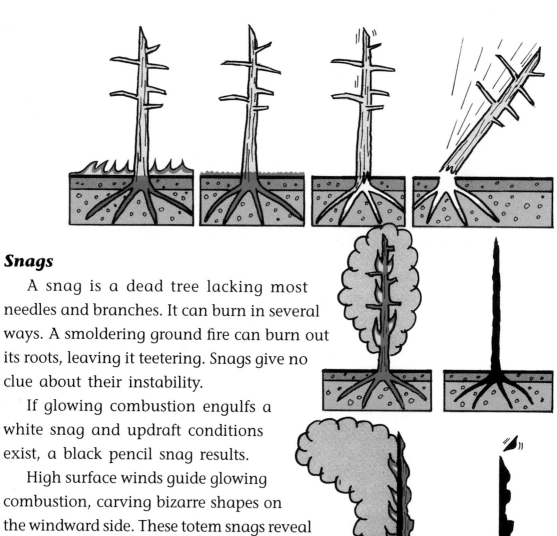

Snags

A snag is a dead tree lacking most needles and branches. It can burn in several ways. A smoldering ground fire can burn out its roots, leaving it teetering. Snags give no clue about their instability.

If glowing combustion engulfs a white snag and updraft conditions exist, a black pencil snag results.

High surface winds guide glowing combustion, carving bizarre shapes on the windward side. These totem snags reveal the wind direction when they burned. The combustion-carved "faces" of totems look toward the wind.

How Wind Affects the Burning of Dead and Living Trees

Surface wind intensifies glowing combustion on the upwind side of snags, producing totems. Conversely, it intensifies flaming combustion on the downwind side of living trees, producing two-tones.

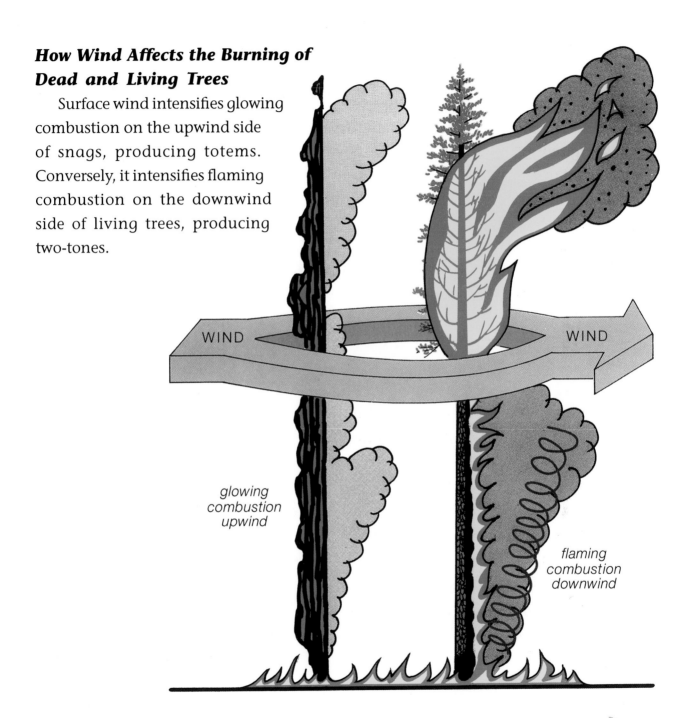

WIND

WIND

glowing combustion upwind

flaming combustion downwind

CONVECTION COLUMNS AND LARGE FIRES

The convection column from a large fire is so powerful that it creates its own local weather.

The parcel of air immediately above a fire consists of very hot, highly energized gas molecules. These energized molecules form a hot, lightweight bubble and push outward and upward into surrounding cooler air. The surrounding ocean of cooler air flows in beneath this hot bubble and lifts it upward into the sky.

The bubbles of hot air above the fire connect to form a continuous column of rising hot gases within a chimney of cooler air. The hot gases rise until they cool to a temperature equal to the surrounding air, which may be greater than 5 miles above the ground.

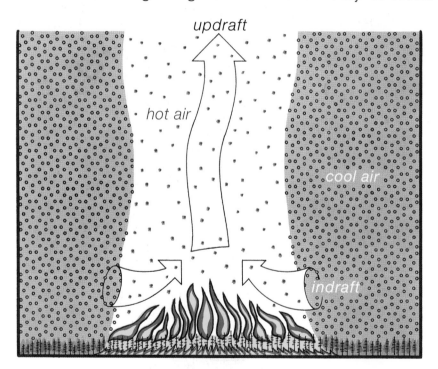

Indrafts of cooler air at a fire's base increase the rate of the combustion of fuels. The updrafts carry smoke, ash, and firebrands high into the atmosphere, where they disperse widely, produce spot fires, and may even alter regional weather.

Plants demonstrate how much sunlight energy they have captured in the producer reaction when they release that energy in a firestorm, nature's most violent and spectacular consumer reaction.

Dramatic changes in fire behavior herald a firestorm: a substantial increase in the fire's rate of combustion and spread, intense radiation, and crowning. The firestorm itself is characterized by a large, violent convection column, very high surface winds, tornado-like fire whirls, and prolific spotting.

The convection column contains hot gases, burning fuels, and a lot of water vapor produced by fuel combustion. At higher altitudes, the water vapor cools and condenses. Condensation forms a fluffy white cumuliform cloud cap from which rain may fall.

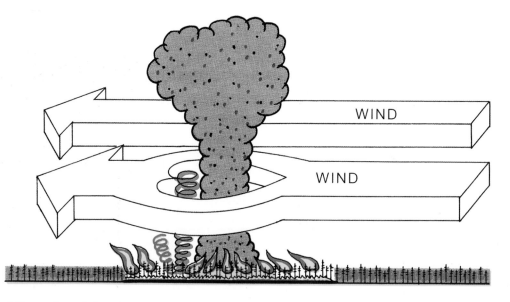

Like a boulder in a stream, strong convection columns act like pillars and force surface winds to diverge around them. The column produces turbulent eddies and whirlpools of air on its downwind side, resulting in dangerous, unpredictable fire behavior.

When winds are strong enough to fracture the convection column, it can result in very rapid fire spread and severe spot fire activity. The fire may eject a rain of firebrands a mile or more downwind, which land at a rate of thousands per acre.

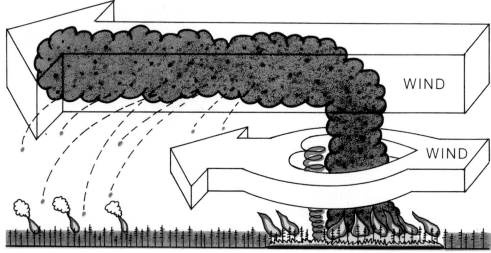

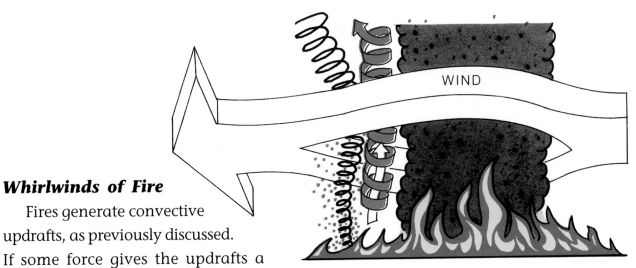

Whirlwinds of Fire

Fires generate convective updrafts, as previously discussed. If some force gives the updrafts a spin, a whirlwind develops. Large fires may produce tornado-like whirlwinds that seem to originate on the turbulent downwind side of the convection column. These are rare, but they can be very destructive.

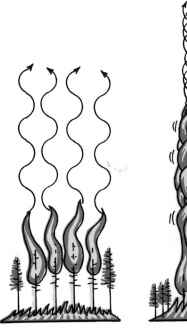

More common are the dustdevil-sized fire whirls that develop from the convection column over intensely burning fuel. Like a twirling ice skater who tucks in her arms to spin faster, the rising column of air suddenly shrinks in diameter and rotates much faster. Burn rates increase. This fire behavior rapidly depletes the fuel and shortens the life span of the fire whirl.

SMOKE

Smoke emitted from a wildland fire consists of two types: gray pyrolysis smoke and black sooty smoke.

Gray pyrolysis smoke originates wherever a cloud of pyrolyzed fuel condenses into tar droplets before it can ignite *(see pages 26–27)*.

Black sooty smoke is especially plentiful in large, turbulent fires. Powerful updrafts rip off bits of flame, which lift skyward. Immense volumes of sooty smoke pour from the openings in torn bags of burning fuel. These rising bubbles of golden flame cool rapidly, and their soot immediately turns from orange to black *(see pages 22–23)*.

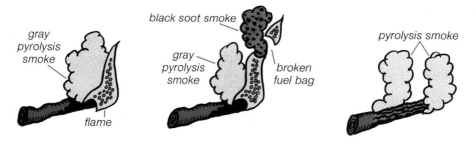

In most wildland fires, the smoke consists of about half tar droplets and half soot and ash particles. If fuels are heavy, wildland fires produce 2 or more tons of smoke particles per acre.

PART IV
Aftermath and Campfire Story

The living arises from the dead, and the dead from the living.

—*Plato*

Wandering about in the ashes of a large fire enlightens the inquisitive mind and piques the imagination. It is like being miniaturized in order to explore the inside of a campfire. The embers and ashes reveal the story of all phases of combustion. They are the remains of an immense chemical reaction that converted millions of tons of fuel to gaseous carbon dioxide and water, producing enormous amounts of heat and light energy. Like magic, solid fuels have changed to gases and floated away.

FIRE CLASSIFICATION SYSTEM

After a wildland fire, forest caretakers analyze its effects on plant and animal life. They determine the severity of soil heating and the amount of plant material killed by the fire. This information guides their predictions about plant regeneration (type of plant and time required), erosion, and stream effects.

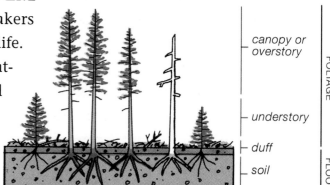

The recovery of plant life on a burn site depends on the depth, intensity, and duration of soil heating. These factors determine the survival of seeds, roots, and underground plant stems. If fire kills these underground plant parts, then natural vegetation recovery will depend on aerial seeding from above and outside the burn site. If the underground plant parts survive, then vegetation resumes growth quickly after the fire. Some green plants may sprout within a few days. Scorched trees that look dead may only be top-killed and will emerge from their roots the following year.

This is a simple fire classification system based on clues found at a fire site.

FOREST FOLIAGE LAYER	partial burn	Mixture of green living trees and black/brown trees that are either top-killed or dead.
	complete burn	Most trees black/brown (dead or top-killed).
FOREST FLOOR LAYER	superficial burn	Blackened forest floor indicates incomplete combustion, superficial soil heating, and high rates of survival for seeds, roots, underground stems, and microorganisms.
	deep burn	White ash or reddish forest floor indicates complete combustion of surface fuels, prolonged deep soil heating, and low survival rates for plant parts and microorganisms.

Flaming combustion of aerial and surface fuels, although dramatic in appearance, minimally heats the forest floor. Most of the searing heat of the fire's reaction zone is directed upwards. A blackened forest floor suggests a superficial burn.

Glowing and smoldering combustion of surface and ground fuels transfer much more heat into the forest floor, and for a very long time. This prolonged, direct energy impulse burns off more of the litter and duff, heats the soil more deeply, and kills more organisms. Prolonged glowing and smoldering combustion produces a deep burn, characterized by white ash and reddish soils.

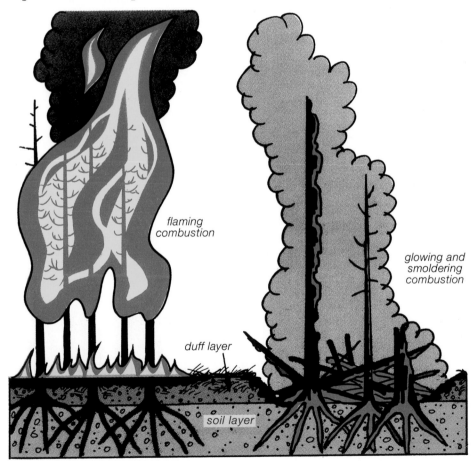

flaming
combustion

glowing and
smoldering
combustion

duff layer

soil layer

MOSAIC BURN

Viewed from a distance, forest foliage burns in a patchwork pattern called a *mosaic*. This irregular burning pattern increases the amount of forest edge, which encourages biological diversity.

A cross section through a mosaic burn site shows areas of differing burn types.

	FOREST FOLIAGE	FOREST FLOOR
1	unburned	unburned
2	unburned	superficial burn
3	partial burn	superficial burn
4	complete burn	superficial and occasional deep burn

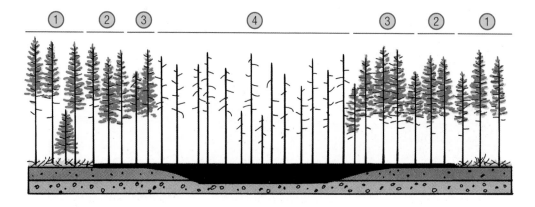

White Lines and Crosses

Dead trees that fell to the ground before the fire may burn up completely because they contain little water. Long after the flames pass, glowing combustion continues to consume their dry trunks. They burn from underside to top because the gray ash falls away from the combustion reaction zone. This prevents the ash from smothering the fire. Patches of rotten wood may smolder for days.

charred duff
deep duff
mineral soil
white ash line
mineral soil with
charred organic matter

The prolonged pulse of heat associated with glowing and smoldering combustion heats the soil deeply and may kill most underground organisms. Complete combustion leaves only white lines of mineral ash.

flaming
combustion

glowing/smoldering
combustion

white ash lines of
complete combustion

Mini-Mosaic

When a grove of trees burns very rapidly, it may produce its own convection column. The column's updrafts and indrafts of air determine the intensity of burn and the direction of flames. During flaming combustion, the convection column is strong. Indrafts direct the flames toward the center of the grove of trees and create a circular burn pattern.

Interior trees, subjected to the most intense heat, burn and char completely. Perimeter trees burn partially, and frequently produce two-tone patterns on tree trunks.

updraft

indraft

indraft

indraft

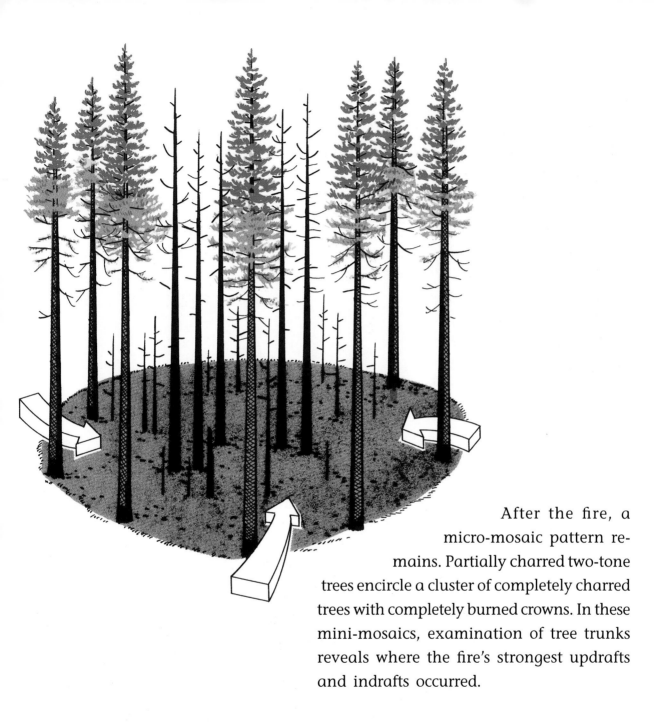

After the fire, a micro-mosaic pattern remains. Partially charred two-tone trees encircle a cluster of completely charred trees with completely burned crowns. In these mini-mosaics, examination of tree trunks reveals where the fire's strongest updrafts and indrafts occurred.

FIRESCAPES

(1) This mature forest scene contains unburned trees, snags, heavy ground fuels, and small young trees.

(2) This firescape displays partially burned trees with dead and scorched needles. Charred litter suggests a superficial burn of the forest floor. White ash lines indicate complete combustion of heavy surface fuels and prolonged, deep heating of the soil. Smoldering fires in root systems of gray snags produce hidden cauldrons of hot coals. *Serious burns to visitors may result if they fall into these cauldrons of coals.*

(3) This firescape reveals the passage of crown fire and surface fire. The trees are thoroughly charred; their crowns were completely burned by intense flaming combustion. The black surface of the forest floor suggests superficial soil heating. Pencil snags developed in windless or updrafting conditions. Numerous white lines and crosses tell of prolonged glowing combustion of heavy surface fuels and deep soil heating beneath the ash. Many snags with burned-out root systems wait for the slightest excuse to fall. These snags kill—without warning!

BURN SITES ARE DANGEROUS, ESPECIALLY ON WINDY DAYS.

(4) A white ash firescape indicates intense fire and complete combustion of aerial and surface fuels. The white remains of incinerated duff and heavy fuels overlie a scorched mineral soil, indicating deep heating of the soil.

(5) High surface winds sculpted this firescape of partially burned two-tone trees and "villages" of totem snags. Tops of totems snap off in the wind. Totems and two-tones are dangerous!

Was the Tree Alive or Dead When It Burned?

Dead, dry, heavy fuel burns by glowing, leaving a shiny black and frequently cobbled surface.

Living, wet, heavy fuel burns by flaming, leaving a dull black surface of charred bark.

dead tree

glowing combustion: *shiny, black, cobbled*

live tree

flaming combustion: *dull black surface*

CAMPFIRE STORY

A teepee of dried pine wood covers a loose pile of equally dry kindling consisting of pine needles and twigs.

The pilot heat dries and preheats the kindling, which signals the pre-ignition phase by its production of tarry smoke. The fuel cloud forms.

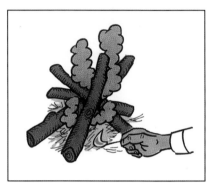

The pilot heat ignites the fuel cloud, which bursts into flames indicating the flaming combustion phase. Kindling flames next preheat the larger fuel to pre-ignition. If all goes well, this kindling produces enough heat to ignite the new fuel cloud formed from pine wood pyrolysis. The pilot heat is no longer needed.

The multishaped fuel clouds ignite as the kindling dwindles to glowing combustion and ash formation.

While the campfire is still fresh, the main body of woody fuel burns primarily by flaming combustion. Pre-ignition, ignition, and flaming combustion phases are easily discernible in different parts of the campfire.

Later on, the campfire exhausts most of its gaseous fuels. Much of the wood has pyrolyzed to char. Glowing combustion burns the char.

Pyrolysis deep inside the larger fuels breaks down the long molecules that provide strength and integrity to the wood. The heat-weakened wood collapses. If heat or oxygen is insufficient to burn these fuels completely, they may smolder, producing tarry smoke.

Gray mineral ash covers the surface of the fuels. It must be constantly knocked off to prevent it from smothering the coals.

The fire goes out, leaving unburned charcoal, a few pieces of unburned wood, and a fluffy pile of mineral ash. Most of the fuel has reacted with oxygen to form carbon dioxide, water, and heat. It's time for bed.

Fire Glossary

For more information on terms used in fire management, look for "FireWords" at www.fire.org, www.firelab.org, or www.frames.gov.

aerial fuel. The standing and supported forest combustibles not in direct contact with the ground, consisting mainly of foliage, twigs, branches, stems, and bark.

backing fire. A fire burning into or against the wind, or down a slope without the aid of wind.

blow-up. A sudden increase in fire intensity and rate of spread, often associated with violent convection.

char. Fuels that are partly broken down and blackened by pyrolysis.

classification system for forest fires

> **forest foliage burn**
>> partial or complete
>
> **forest floor burn**
>> *low-severity burn.* A degree of burn that leaves the soil covered with partially charred organic material; large fuels are not deeply charred. (Comparable to human first-degree burn.)
>>
>> *moderate-severity burn.* Degree of burn in which all organic material is burned away from the surface of the soil layer, which is not discolored by heat. Any remaining fuel is deeply charred. Organic matter remains in the soil immediately below the surface. (Comparable to human second-degree burn.)

> *severe burn.* Degree of burn in which all organic material is burned from the soil surface, which is discolored by heat (usually red). Organic material below the surface is consumed or charred. (Comparable to human third-degree burn.)

classification system for soil heating

> *low.* Heating of mineral soil insignificant.
>
> *moderate.* Up to 5 centimeters of mineral soil charred.
>
> *high.* Greater than 10 centimeters of mineral soil charred and surface often reddish due to mineral oxidation.
>
> *severe.* Soil crystallized to a hardened surface of larger mineral crystals.

combustion. Consumption of fuels by oxidation, producing heat, generally flame, and/or incandescence.

combustion reaction zone. Walls of the flame bag, where pyrolyzed fuel gases and oxygen combine to release the energy originally acquired from the sun.

conduction. Transfer of heat from atom to atom or molecule to molecule of a solid substance; energy travels from a region of higher temperature to a region of lower temperature.

consumer reaction. Disassembly of the products of photosynthesis to obtain sunlight energy stored in their chemical bonds.

convection. Transfer of heat by the movement of a gas or liquid. In meteorology, vertical transfer of heat in the absence of wind.

crown. The upper part of a tree or other woody plant, carrying the main branch system and foliage.

crown fire. A fire that advances from top to top of trees or shrubs. Sometimes crown fires are classed as either independent or dependent, to distinguish their degree of independence from the surface fire.

dead fuel. Fuels having no living tissue. The moisture content is governed almost entirely by atmospheric moisture (relative humidity and precipitation), air temperature, and solar radiation.

decomposition. The physical breakdown of complex materials (organic or inorganic) into constituent parts.

duff. Forest floor material composed of fuels in various stages of decomposition. Includes the decomposing litter or fermentation (F) layer, and the organic soil or humus (H) layer.

fine fuels/flash fuels. Fuels, such as grass, leaves, draped pine needles, ferns, tree moss, and some slash, that ignite readily and burn rapidly.

firebrand. Any burning material, such as leaves, wood, and glowing charcoal or sparks, that could start a forest fire.

firestorm. Violent convection caused by a large, continuous, intense fire. Often characterized by violent, surface indrafts; a towering convection column; long-distance spotting; and sometimes tornado-like vortices.

fire triangle. An instructional aid in which the sides of a triangle are used to represent the three factors (oxygen, heat, and fuel) necessary for combustion. When any one of these factors is removed, combustion ceases.

fire whirl. A spinning column of ascending hot air and gases rising from a fire and carrying aloft smoke, debris, and flame. Fire whirls range from 1 to 2 feet in diameter to small tornados in size and intensity. They may involve the entire burning area or only a hot spot within the fire area.

flame. A mass of gas undergoing rapid combustion, generally accompanied by the production of heat and incandescence. The combustion is contained within the **flame bag**, or **fuel bag**, the outer edge of which is defined by the combustion reaction zone.

flaming combustion. Luminous oxidation of the gases evolved from pyrolysis of fuel.

flaming front. That zone of a moving fire where the combustion is primarily flaming. Behind this flaming zone combustion is primarily glowing and smoldering.

flanking fire. That part of a fire that is roughly perpendicular to the main direction of the heading and backing fires.

forest. Generally, an ecosystem characterized by a more or less dense and extensive tree cover. More particularly, a plant community predominantly of trees and other woody vegetation growing more or less closely together, and composed of an overstory (canopy) and understory.

forest floor. The duff and soil layers supporting the forest.

fuel. Combustible material. *See also* aerial fuel; dead fuel; fine fuel; ground fuel; heavy fuel

fuel cloud. A cloud of flammable gas molecules such as fat, oil, turpene, alcohol, and resin, and pyrolyzed fuel molecules that are released from wood during heating.

glowing combustion. Oxidation of gases evolved from fuel within the matrix of charred wood, producing incandescence but little or no flame.

ground fire. Fire that burns organic material in the duff and mineral soil layers.

ground fuel. All combustible materials below the surface litter, including duff, tree or shrub roots, punky wood, peat, and sawdust. Ground fuels normally support glowing combustion without flame and smoldering combustion.

heading fire. A fire spreading with the wind or uphill.

heat transfer. The process by which heat energy is imparted from one body to another through conduction, convection, or radiation.

heavy fuels. Fuels of large diameter, such as snags, logs, and the wood of large branches, or of a peaty nature, that ignite and burn more slowly than flash fuels.

indraft. Cool air moving horizontally toward a fire.

ladder fuels. Fuels that provide vertical continuity between strata. In ladder fuels, fire is able to carry from surface fuels into crowns with relative ease.

litter. Dead material in the top layer of the forest floor. It is composed of loose debris of dead sticks, branches, twigs, and recently fallen leaves or needles little altered in structure by decomposition.

living fuels. Naturally occurring fuels in which the moisture content is physiologically controlled within the living plant.

low-severity burn. See **classification system for forest fires.**

mineral soil. Soil layers below the predominantly organic soils of the duff layer. Mineral soil contains few or no combustibles.

moderate-severity burn. See **classification system for forest fires.**

parts of a fire. In typical free-burning fires the spread is uneven, with the main spread moving with the wind or upslope. The most rapidly moving portion is designated the **head** of the fire, the adjoining portions of the perimeter at right angles to the head are known as the **flanks,** and the slowest moving portion is known as the **rear,** or **back.**

pilot heat. External source of heat needed to start combustion.

producer reaction. Photosynthesis, the process by which plant life converts sunlight energy into food, a useable and storable form of chemical energy.

pyrolysis. The decomposition of fuel at an elevated temperature.

radiation. Propagation of energy through space.

reaction zone. See **combustion reaction zone.**

scorch. Permanent discoloration of plant parts caused by heating, which may be black if charred, or yellowish brown if killed by heating but not charred or burned.

severe burn. See **classification system for forest fires.**

smoldering combustion. Flameless combustion in a fuel-rich and very oxygen-poor environment within stumps, dead logs, underground roots, or other tightly packed fuels.

snag. A standing dead tree or standing portion from which at least the leaves or needles and smaller branches have fallen. Often called a stub if less than 6 meters tall.

soot. Carbon from the incomplete combustion of wood.

spot fire. Fire set outside the perimeter of the main fire by firebrands.

surface fire. Fire that burns only surface litter, other debris on the forest floor, and surface vegetation.

surface fuel. Combustibles on the forest floor, normally consisting of fallen leaves or needles, twigs, bark, cones, and small branches that have not yet decayed sufficiently to lose their identity. Also grasses, low shrubs, small trees, heavier branches, downed logs, stumps, seedlings, and forbs.

tar. Droplets of pyrolyzed fuels that have not burned completely.

top-killed. Used to describe plants with above-ground tissues killed by fire, but with below-ground tissues (such as roots and underground stems) still alive and able to resume growth.

torching. A tree (or small clump of trees) is said to "torch" when its foliage ignites and flares up, usually from bottom to top.

two-tone. An unevenly burned tree trunk—the downwind side of the trunk is burned more severely than the upwind side.

updraft. Hot air moving up in a column above a fire.

volatiles. Readily vaporized organic materials that, when mixed with oxygen, are easily ignited.

wildland. An area in which development is essentially nonexistent, except for roads, railroads, and powerlines. Structures, if any, are widely scattered, and are primarily for recreational purposes.

wildland fire. Any fire occurring in wildland except a prescribed fire.

Recommended Reading

Arno, Stephen F., and Stephen Allison-Bunnell. *Flames in Our Forest—Disaster or Renewal?* Washington, D.C.: Island Press, 2002.

Brown, Arthur Allen. *Forest Fire: Control and Use.* New York: McGraw-Hill, 1973.

Chandler, C., P. Cheney, P. Thomas, L. Trabaud, and D. Williams. *Fire in Forestry, Vols. I, II.* New York: John Wiley and Sons, 1983.

Cone, Patrick. *Wildfire.* Minneapolis, Minnesota: Carolrhoda Books, 1997.

DeBano, Leonard F., Daniel G. Neary, and Peter F. Ffolliott. *Fire's Effects on Ecosystems.* New York: John Wiley and Sons, 1998.

Johnson, Edward A., and Kiyoko Mianishi. *Forest Fires: Behavior and Ecological Effects.* San Diego: Academic Press, 2001.

Pyne, Stephen J., Patricia L. Andrews, and Richard D. Laven. *Introduction to Wildland Fire, 2nd edition.* New York: John Wiley and Sons, 1996.

Pyne, Stephen J., and William Cronon. *Fire in America: A Cultural History of Wildland and Rural Fire.* Seattle: University of Washington Press, 1997.

Pyne, Stephen J., and William Cronon. *World Fire: The Culture of Fire on Earth.* Seattle: University of Washington Press, 1997.

Whelan, Robert J. *The Ecology of Fire.* New York: Cambridge University Press, 1995.

Wolf, Thomas J. *In Fire's Way: A Practical Guide to Life in the Wildfire Danger Zone.* Albuquerque: University of New Mexico Press, 2003.

Recommended Internet Sites

How to reduce the risk that wildland fire will destroy your home:
http://www.firewise.org/

Science of wildland fire, presented on Smokey Bear's Web site:
http://www.smokeybear.com/science.asp

Life After Fire: How Fire Affects Wildland Plants and Animals on the Lolo National Forest, western Montana:
http://www.fs.fed.us/r1/lolo/wl-fire-ecology/fire1.html

Messages from land management agencies about fire management and fire's role in wildlands:
http://www.fs.fed.us/fire/fireuse/wildland_fire_use/role/role_pg1.html

Maps depicting fire weather, vegetation greenness, and fire danger in the U.S.—updated regularly:
http://www.fs.fed.us/land/wfas/

Summaries of scientific findings about the ecology and relationships of 1,000 plant and animal species to fire:
http://www.fs.fed.us/database/feis/

Maps of historic fire regimes and current vegetation condition throughout the U.S.:
http://www.fs.fed.us/fire/fuelman/

Current research on fire behavior, fire effects, and the chemistry and dispersion of smoke from wildland fires:
http://firelab.org

For the latest information about fires burning during fire season, link to the national Interagency Fire Center:
http://www.nifc.gov/information.html

Index

71

Acknowledgments

The following individuals, listed alphabetically, contributed their valuable time and intellect to the production of this manuscript:

Don Despain, Park Biologist in the Research Division, Yellowstone National Park. Thank you for going out into the firescapes and for manuscript critique.

Bryan Jenkins, Ph.D., Associate Professor, Agricultural Engineering Department, University of California at Davis. Thank you for knowledge about cellulose combustion.

Ian Kennedy, Ph.D., Associate Professor, Chemical Engineering Department, University of California at Davis. Thank you for the long hours of questions and answers about flame and gas combustion.

George B. Robinson, Chief of Interpretation, Yellowstone National Park. Thank you for recognizing the need for general knowledge about fire science and telling me about it—for all your efforts in fund-raising, for your manuscript critique, and for your home phone number.

Richard Rothermel, Research Engineer and Project Leader (retired) at the Fire Sciences Laboratory, Missoula, Montana. Thank you for your immensely invaluable red-pencil criticism and positive support toward scientific accuracy.

Henry Shovik, Ph.D., Soils Scientist, Yellowstone National Park/Gallatin National Forest. Thanks for your positive support and knowledge about soils and forest regeneration after fire.

Mary Jane Sligar, M.S., "The non-scientist point of view." Thank you for clear-thinking observations and editing.

Darold Ward, Ph.D., Supervisory Chemist and Project Leader (retired) at the Fire Sciences Laboratory, Missoula, Montana. Thank you for your critical review of the manuscript.

Lee, wherever you are.

Mountain Press would like to thank the following experts who provided technical updates to the second edition:

Patricia L. Andrews, Research Physical Scientist, Fire Sciences Laboratory, Missoula, Montana

Roberta Bartlette, Forester, Fire Sciences Laboratory, Missoula, Montana

Bret Butler, Research Mechanical Engineer, Fire Sciences Laboratory, Missoula, Montana

Garon C. Smith, Professor of Chemistry, The University of Montana, Missoula, Montana

Jane Kapler Smith, Ecologist, Fire Sciences Laboratory, Missoula, Montana

Paul A. Werth, Meteorologist, Northwest Coordination Center, Portland, Oregon

William H. Cottrell Jr. spent his childhood exploring life in and around a farm pond in Missouri. Colorado State University granted him a B.S. in zoology, and he attended medical school at the University of Missouri. During the Vietnam era, he served three years with the U.S. Army in Alaska. His formal education concluded with an orthopedic surgical residency at the University of Southern California.

A private surgical practice, a family, and a passion for rock climbing took up much of his time until a climbing accident temporarily slowed him down. This provided Cottrell the opportunity to learn how to fly and to develop and publish AIDS-related educational programs, the first edition of *The Book of Fire,* and three rock climbing guidebooks. His ability to present complex scientific information by combining accurate graphics and simplified text grew out of the doctor-patient relationship. He continues to practice medicine as director of the Osteoporosis Medical Center in Cameron Park, California. He frequently lectures on the molecular biology of bone and bone disease, and still loves to rock climb, free-heel ski, and fly in storms.